YOUR BODY

Written by Clare Trotman

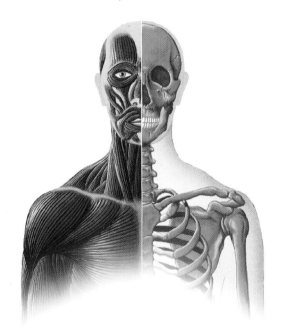

TOP THAT! Kids™

Contents

Introduction

Take a good look at yourself in the mirror. It may surprise you to learn that what you see is far more complex and impressive than anything else on Earth. In fact, you're amazing!

Fascinating Facts

From the brain to the senses, the heart, or even the hair that grows on your head, most people want to learn a little bit more about the super structure that is the human body.

Body Bits

With the help of some amazing images, this book tackles some of the big questions concerning your body and what goes on inside it. In fact, it will even give you answers to smaller questions which you may never have thought of asking!

Glossary Guidance

If you forget what a word means, check out the glossary at the back of this book. It contains explanations of most of the terms used, and is there for you to refer to whenever anything needs clarifying. So are you ready? Then get set for a fascinating journey around the human body!

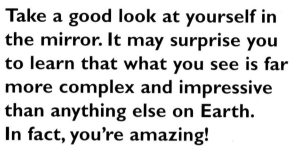

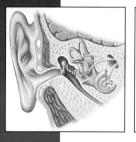

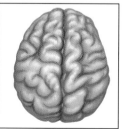

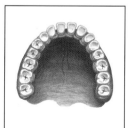

The brain is your body's control center. It sits within the skull and is more powerful than any computer. The brain is made up of billions of cells, which send and receive messages from the body via the spinal cord and nerves.

Cerebrum

The cerebrum is the largest part of the brain and controls movement, thoughts, sensations, and emotions. It is divided into two parts, the right and left cerebral hemispheres. The right side of the brain controls the movements for the left side of the body. The left side of the brain controls the movements for the right side of the body.

Brain Stem & Cerebellum

The rest of the brain is found under the cerebrum and consists of the brain stem and cerebellum. The cerebellum co-ordinates movement, posture, and balance. The brain stem (medulla) controls functions such as heart rate, breathing and coughing.

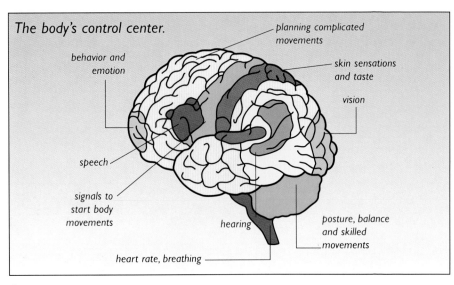

The body's control center.

planning complicated movements

behavior and emotion

skin sensations and taste

vision

speech

signals to start body movements

hearing

posture, balance and skilled movements

heart rate, breathing

Body Facts

• On average, we only ever use approximately a third of our brain. Imagine what we could do if we used all of it!

• The average adult brain weighs over 3 pounds!

How the brain is made up.

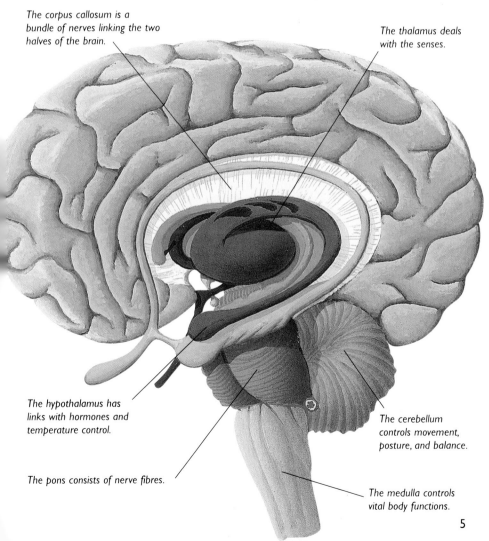

The corpus callosum is a bundle of nerves linking the two halves of the brain.

The thalamus deals with the senses.

The hypothalamus has links with hormones and temperature control.

The cerebellum controls movement, posture, and balance.

The pons consists of nerve fibres.

The medulla controls vital body functions.

5

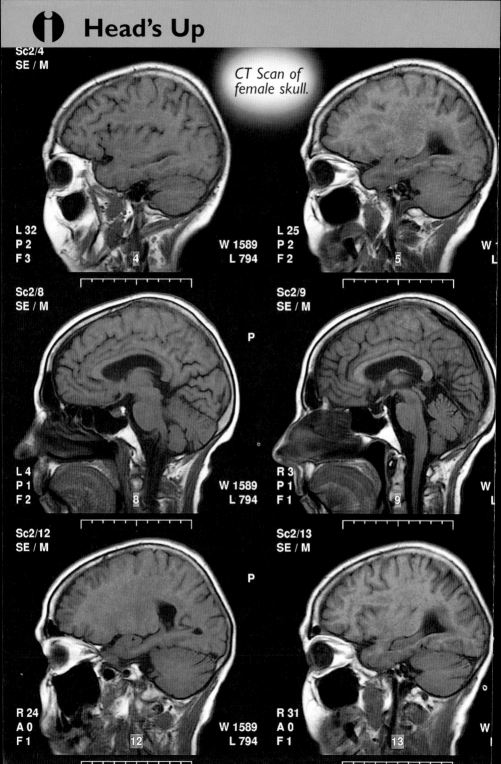

CT Scan of female skull.

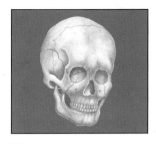

The head contains the major organs, which allow you to see, hear, smell, and taste. It also contains your brain, and the bones which make up your face.

Skull
The skull consists of 22 bones, eight of which protect the brain, and fourteen that form the face. It is one of the most important structures in your body as it provides a safe home for the fragile brain and sensory organs.

Mandible
The mandible, or jaw, is the only moveable bone in the skull. If it didn't move, you wouldn't be able to eat or talk properly.

Eyes
The eyes sit deep within the eye sockets, to protect them from injury.

Nose
The nose is mainly made up of cartilage and is not considered part of the bony skull.

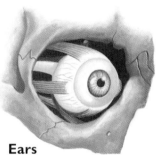

Ears
The majority of the structures that form the ear are protected within the skull. The only visible part of the ear is the pinna.

An X-ray image of the skull.

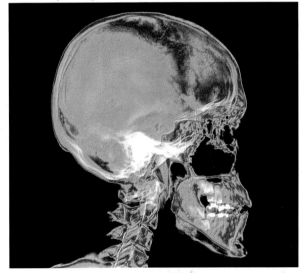

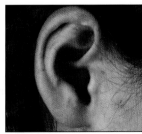

Teeth

Teeth have different functions depending on their position, shape, and size. Incisors and canine teeth allow us to bite off pieces of food. Molars are used for grinding and chewing. Most children have 20 "milk" teeth, while adults have 32 permanent teeth.

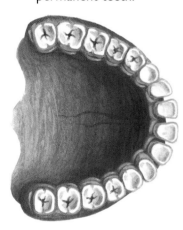

Pharynx

The pharynx, or throat, is 4-6 inches long, and lies behind the nose, mouth and larynx (voice box). Food passes from the mouth into the throat, and then into the esophagus (gullet).

Facial Muscles

There are over 100 different muscles in the face. They enable us to speak, eat, laugh, and cry.

Tongue

The tongue is a muscle that helps us to chew, swallow, speak, and taste. It is a little-known fact that tongue rolling is a genetic trait. This means you can either roll your tongue or you can't, but it is not something you can learn to do. Can you roll your tongue?

Larynx

The larynx is your voice box. The sound of your voice depends upon the length of your vocal cords and the position of your

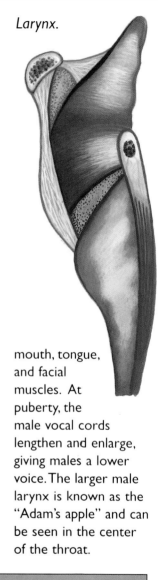

Larynx.

mouth, tongue, and facial muscles. At puberty, the male vocal cords lengthen and enlarge, giving males a lower voice. The larger male larynx is known as the "Adam's apple" and can be seen in the center of the throat.

Body Facts

- It takes 30 muscles to smile and 40 to frown—so be happy to look years younger!

- Everyone has a unique tongue print, just as everyone has a unique finger print.

Cross-section of the head.

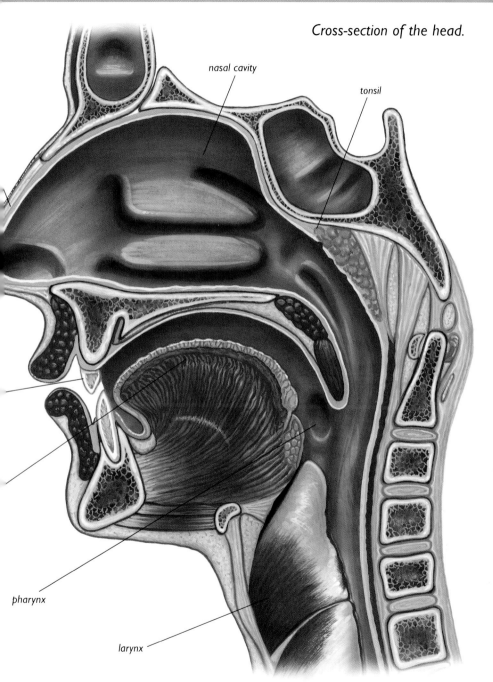

nasal cavity

tonsil

pharynx

larynx

Sight and Sound

The Eye

The eye is ball-shaped and approximately one inch wide. Six muscles attach the eye to the eye socket and allow it to move.

Light waves enter the eye through the pupil. The cornea and lens help focus the light onto the retina, at the back of the eye. The retina then sends a message to our brain via the optic nerve, allowing us to see.

cornea

iris

optic nerve

pupil

lens

retina

In bright light our pupils get smaller and in the dark they get bigger, trying to absorb more light. The eye is able to focus on objects that are near or far by changing the thickness of the lens. Your ability to see objects and perspective are affected when you are only able to see out of one eye. Try covering one eye and see how the distance and position of an object will appear to change.

Body Facts

- About 10% of men and 0.5% of women are color blind. The most common form of this makes it hard to distinguish between red, brown, and green.

- An image reflected onto the retina is actually upside down. Our brain flips it so we see it the way it is.

The Ear

The ear is made up of three sections: the outer, middle, and inner ear. Vibrations in the air, known as sound waves, enter the outer ear and hit the eardrum. This sends vibrations into the middle ear, through the ossicles, and causes the cochlea (inner ear) to vibrate. This stimulates the auditory nerve to send sound to our brain and enable us to hear.

How the ear is made up.

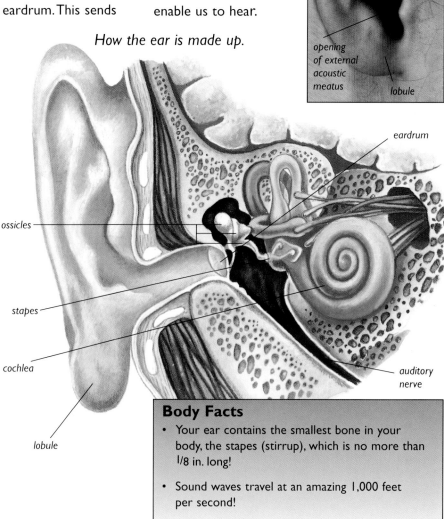

helix

opening of external acoustic meatus

lobule

eardrum

ossicles

stapes

cochlea

lobule

auditory nerve

Body Facts

- Your ear contains the smallest bone in your body, the stapes (stirrup), which is no more than 1/8 in. long!

- Sound waves travel at an amazing 1,000 feet per second!

Taste, Touch, and Smell

The Tongue

The tongue contains approximately 10,000 taste buds and is able to detect four types of tastes: bitter, sweet, sour, and salty. Taste is also detected by taste buds on the palate, roof of the mouth, and throat.

The tongue is sensitive to heat, cold, and pain. Your sense of taste can be affected if you are unwell, as it is linked closely with smell.

You can taste hot foods more easily than you can cold foods. Try eating the same food hot and cold and see if you can taste the difference.

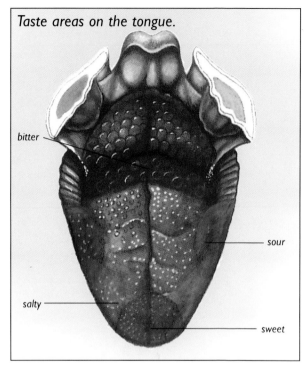

Taste areas on the tongue.

bitter

sour

salty

sweet

The Nose

The nose has two functions: smelling and breathing. Millions of olfactory receptor cells in the nose connect with the brain via the olfactory nerve, enabling us to smell. Sniffing increases our sense of smell, because it increases the speed and volume of airflow into the nose.

Touch

The skin contains nerve endings, which are stimulated by pain, heat, cold, and touch. Our fingertips, lips, and tongue are some of the most sensitive parts of our body.

Body Facts

- By the time you are 60, you will have lost half of your taste buds!

- You can only smell one odor at a time.

- Prolonged exposure to one smell will reduce your awareness of it. This is probably why some people do not realize that they have smelly feet!

Our sense of smell.

Olfactory nerve
endings and nerves.
These sense smells.

The olfactory bulb.
This sends smells
to the brain.

Olfactory tract.
This carries messages
received by the olfactory
bulb to the brain.

Nostril.
Smells enter the
nose here.

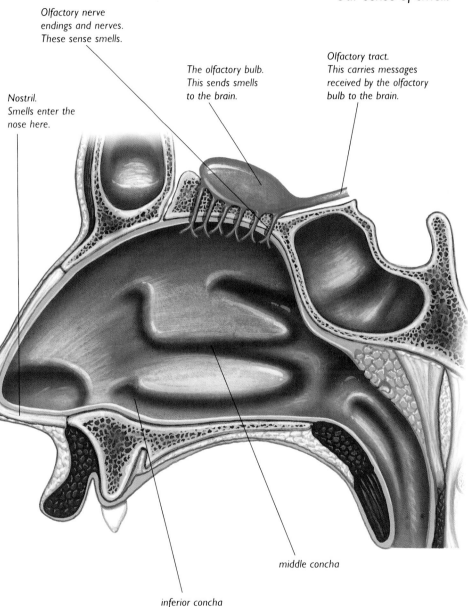

middle concha

inferior concha

Beautiful Bones

The skeleton.

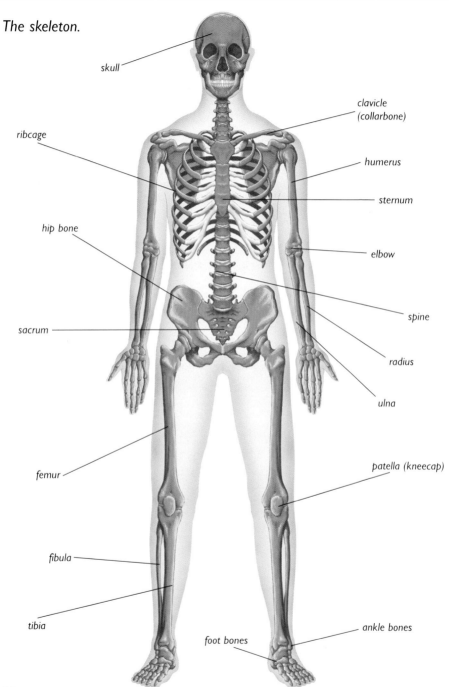

skull

clavicle (collarbone)

ribcage

humerus

sternum

hip bone

elbow

sacrum

spine

radius

ulna

femur

patella (kneecap)

fibula

tibia

ankle bones

foot bones

The adult skeleton is made up of 206 bones. The skeleton protects the major body organs and allows us to stand upright. Without a skeleton, we would be like enormous jellyfish!

The skeleton can be divided into central bones, the skull, spine, sternum and ribs, and bones that form our shoulders, arms, pelvis, and legs.

Bone Make-up

Bones are made up of 20% water, 30-40% "organic" material, such as bone cells, and 40-50% "inorganic" material, such as calcium. Bones provide an anchor for muscles, tendons and ligaments to attach themselves to, allowing movement. They contain bone marrow, which makes blood cells.

Movement

The hip and shoulder are ball-and-socket joints, which allow movement in several directions. The knee and elbow are hinge joints, which only allow movement in one plane, for example forwards and backwards. All joints are covered with cartilage, which helps to protect them from wear and tear.

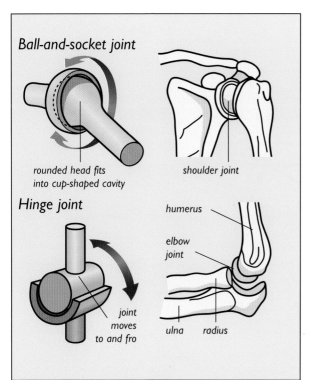

Ball-and-socket joint

rounded head fits into cup-shaped cavity

shoulder joint

Hinge joint

joint moves to and fro

humerus

elbow joint

ulna   radius

Body Facts

- Babies are born with around 300 bones, which fuse together over time to create a stronger, larger skeleton.

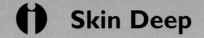

The largest organ in the human body is your skin. On average, your skin has a surface area of over 20 square feet. Our skin protects us from the Sun's ultraviolet rays, harmful bacteria, and prevents us from losing too much water. The skin also controls our temperature, allows us to touch, forms vitamin D, and secretes waste substances like salts.

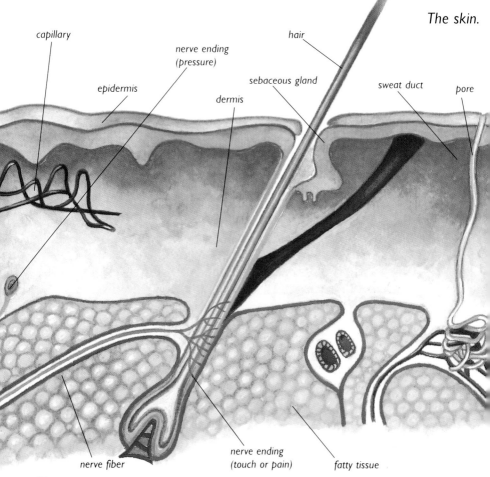

The skin.

capillary

nerve ending
(pressure)

hair

epidermis

sebaceous gland

sweat duct

pore

dermis

nerve fiber

nerve ending
(touch or pain)

fatty tissue

Skin Deep

Body Facts

- We lose around twelve million skin cells every day!

- Every square inch of skin contains over 20 feet of blood vessels and around 30 million bacteria!

- Nails are made of keratin (a tough protein) and protect the tips of our fingers and toes.

Cross-section of finger and nail.

- Fingernails grow three to four times faster than toenails, and both grow faster in hot temperatures.

- A typical human scalp contains 100,000 hairs. Comprised mainly of a substance called keratin, only the base of the hair, where it grows, is alive.

- A nail takes, on average, six months to grow from base to tip.

- Hair grows at a rate of around half an inch per month. Beards are the fastest-growing hairs on the human body.

Cross-section of hair in the skin.

- If the average male did not shave, his beard would grow to 30 feet long in a lifetime.

Skin Layers

Skin is made up of two layers—the outer epidermis and the inner dermis.

The epidermis is constantly growing and is covered by a layer of dead cells which are constantly being shed by our bodies.

The dermis contains over three million "sweat glands," hair follicles, sensory receptors, nerves, and capillaries (blood vessels).

It is four times thicker than the epidermis and contains a lot of collagen, which helps repair any cuts.

Hot and Cold

If we get too hot, our skin cools us down by sweating and vasodilating (widening the blood vessels). If we get too cold, our skin warms us up by shivering and vasoconstricting (narrowing the blood vessels), in an attempt to keep heat in.

There are three types of muscle in our body, skeletal, cardiac, and smooth. We only have conscious control over our skeletal muscle.

Skeletal muscles anchor to bones or other muscles via a tendon. Cardiac muscle makes up the muscle of our heart. Smooth muscle lines our organs, such as the bladder and blood vessels.

Skeletal Muscle

Skeletal muscle is the main muscle type in our bodies and makes our bones move by contracting (shortening) and relaxing (lengthening). Skeletal muscle is sometimes called voluntary muscle. Reflex actions are produced by the skeletal muscles working with bones and tendons.

600 Varieties

There are over 600 different muscles in our body. Each one has a different shape, size, and function.

Keep it There!

Muscles help to keep our abdominal organs in place.

Ligaments

Ligaments attach bone to bone, preventing unwanted movement around joints.

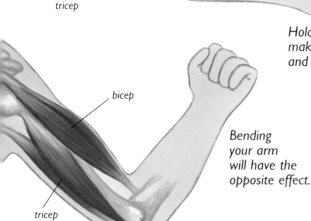

bicep

tricep

bicep

tricep

Holding your arm out will make the bicep muscle extend and the tricep muscle contract.

Bending your arm will have the opposite effect.

Body Facts

- Gluteus maximus (the bottom muscle to you and me) is the largest and most powerful muscle in your body.

The muscle network.

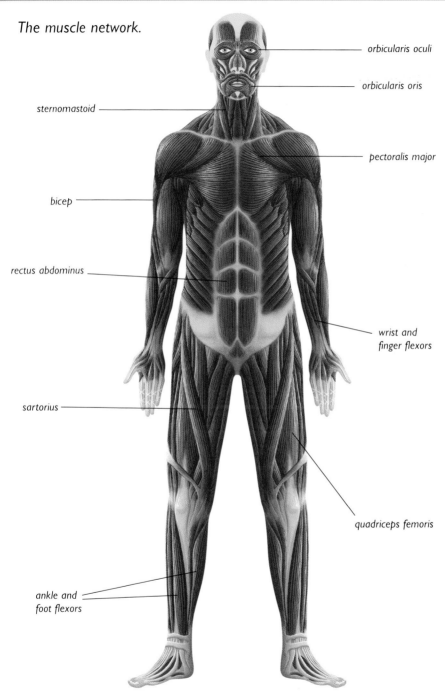

orbicularis oculi

orbicularis oris

sternomastoid

pectoralis major

bicep

rectus abdominus

wrist and
finger flexors

sartorius

quadriceps femoris

ankle and
foot flexors

The nervous system includes the brain, spinal cord, and nerves. It is divided into three areas, the central nervous system (brain and spinal cord), the peripheral nervous system (31 pairs of spinal nerves), and the autonomic nervous system (twelve pairs of nerves, which connect directly to the brain).

In a Cell

A single nerve contains millions of nerve cells. Each nerve cell consists of an axon (a group of fibers) connected to a neuron (the nerve cell body). The neuron has several finger-like branches called dendrites which allow signals to be passed between the nerve cells.

System Control

The nervous system is probably the most important of the body's systems. It is vital for the development of language, thought, and memory, it controls movement, and it regulates breathing. All the body's nerves are ultimately controlled by the brain and the spinal cord.

Nerve Types

There are two main types of nerves, sensory and motor. Sensory nerves carry information from our sensory organs, like the eyes and skin, to our brain.

Motor nerves carry information from our brain to our muscles, making them move.

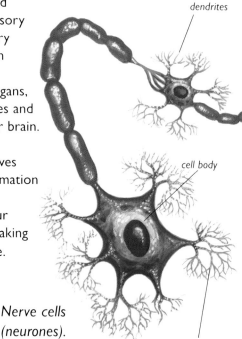

dendrites

cell body

dendrites

Nerve cells (neurones).

The nervous system.

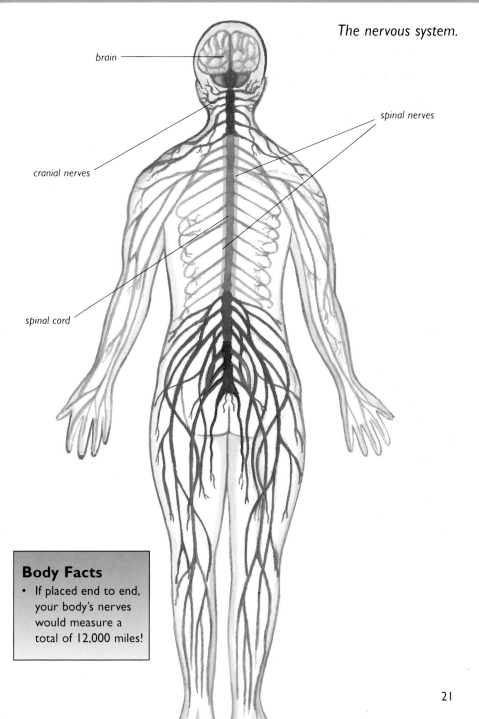

brain

cranial nerves

spinal nerves

spinal cord

Body Facts

- If placed end to end, your body's nerves would measure a total of 12,000 miles!

21

How the respiratory
system is made up.

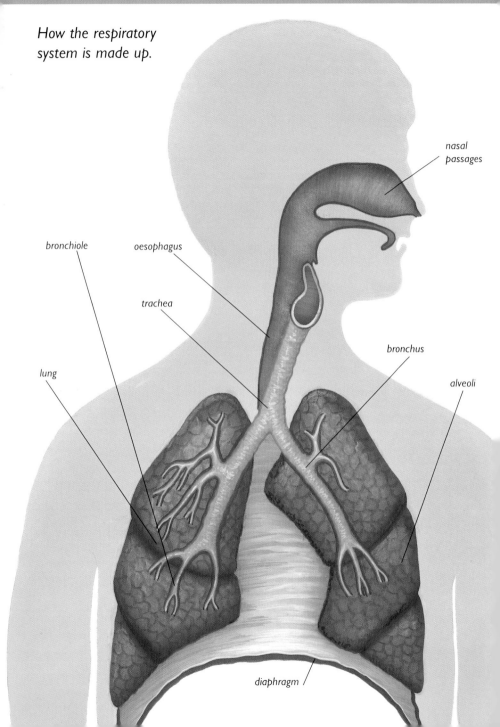

The human body needs a constant supply of oxygen to stay alive.

Inspiration!
As we breathe in (inspiration), air enters the body through the nose or mouth and travels down the trachea (windpipe), where it divides between two bronchi.

In and Out
The air then passes through our lungs via thousands of bronchioles (tiny tubes) and into the millions of alveoli (air sacks), allowing oxygen to pass into our blood. As we breathe out (expiration) we get rid of carbon dioxide, a waste product of respiration. The ribs, muscles around the ribs and diaphragm help our lungs to fill with air. If you take a deep breath you will notice that your chest gets bigger and as you breathe out it gets smaller.

Big Breaths
Every minute we breathe in around 400 cubic inches of air—more if we are exercising. That is enough air to blow up three small balloons!

Breathing Rate
The average adult breathing rate is 12-15 breaths per minute. This rate will increase if we are exercising and decrease when we are resting.

Regulating Breathing
The rate at which we breathe is controlled by the brain and regulated by the level of carbon dioxide in the blood. When exercising, we take in more carbon dioxide, so the brain tells us to take deeper breaths so that more oxygen is inhaled. This makes the heartbeat increase, so that the carbon dioxide is more quickly burned off.

Holding Your Breath
If you hold your breath, no oxygen or carbon dioxide enters the respiratory system and so, very soon, your brain stops receiving any stimulation. This can cause you to lose consciousness.

Body Facts
- The surface area of our lungs is over 1,000 square feet!

The heart is about the size of a tennis ball, weighs around nine ounces, and sits in the middle of your chest, slightly to the left. It is a powerful muscle, and constantly pumps blood around the body, never getting tired.

Blood Vessels

Arteries carry oxygenated blood from the heart to the tissues and organs. Veins carry deoxygenated blood from the tissues and organs back to the heart.

Blood

Blood is made up of red cells, white cells, and platelets, all floating in a liquid called plasma. The red blood cells contain a pigment called hemoglobin that carries oxygen to the tissues. The white blood cells help fight infection and the platelets help our blood to clot if we are injured.

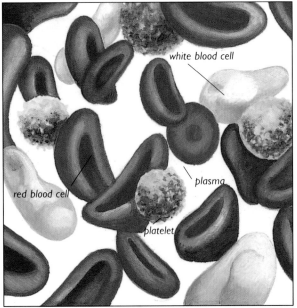

white blood cell

red blood cell

plasma

platelet

The make-up of blood.

The heart is made up of four chambers, the right and left atria and the right and left ventricles. The right side of the heart pumps blood to our lungs. The left side of the heart pumps blood around the rest of the body.

Body Facts

- The heart pumps 2,500-4,000 gallons of blood around the body every day.

- A single drop of blood contains over ten million cells!

The heart.

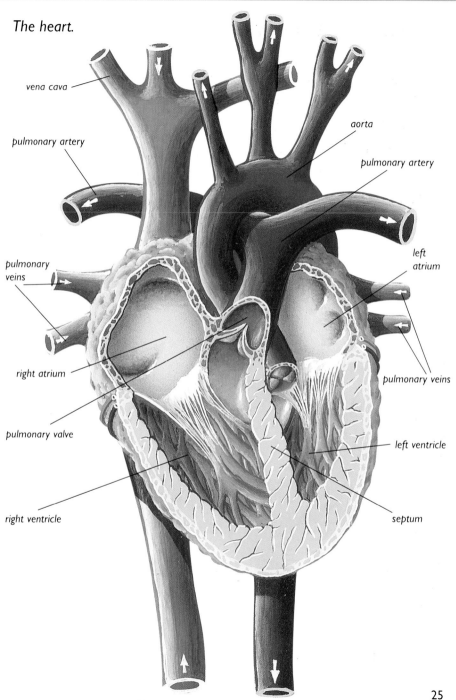

vena cava

pulmonary artery

aorta

pulmonary artery

pulmonary veins

left atrium

right atrium

pulmonary veins

pulmonary valve

left ventricle

right ventricle

septum

We eat to give us the energy our bodies need to function. The process of converting food into energy is known as digestion.

Passing Through

Food enters the mouth where it is chewed and mixed with saliva, allowing us to swallow. It then passes down a tube called the esophagus and into the stomach. From the stomach, it passes to the small and then large intestine, before ending its journey in the rectum. It takes around 24 hours for food to pass through our digestive system.

Breaking it Down

The liver, gall bladder, and pancreas release special chemicals called enzymes, which help break down food.

Food Processing

The foods that take the longest to process are fatty foods such as pies and cakes while starch-based carbohydrates such as potatoes and bread are processed the quickest. Digestion usually takes between two and six hours.

Feeling Hungry?

The stomach expands as it is filled with food. The food is then mixed with acids and enzymes before being squeezed through the small intestine. The stomach then contracts and it is at about this time when you may begin to feel hungry again.

Upset Tummy

If you eat something that hasn't been properly cooked, it may well contain harmful bacteria. The body will try to get rid of the bacteria as quickly as possible by either rushing it through the system or by relaxing the muscles and sending the food back up so that you vomit.

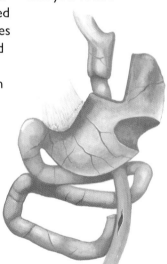

The stomach.

Body Facts

- The stomach actually sits just under the ribs, not around your belly button.

- The stomach can hold half a gallon of digested food.

- We produce 30-60 ounces of saliva a day.

The digestive system.

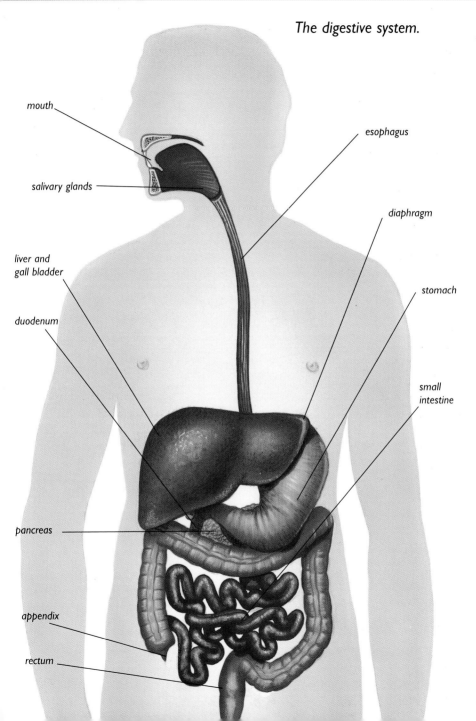

mouth

esophagus

salivary glands

diaphragm

liver and
gall bladder

stomach

duodenum

small
intestine

pancreas

appendix

rectum

The waste management system.

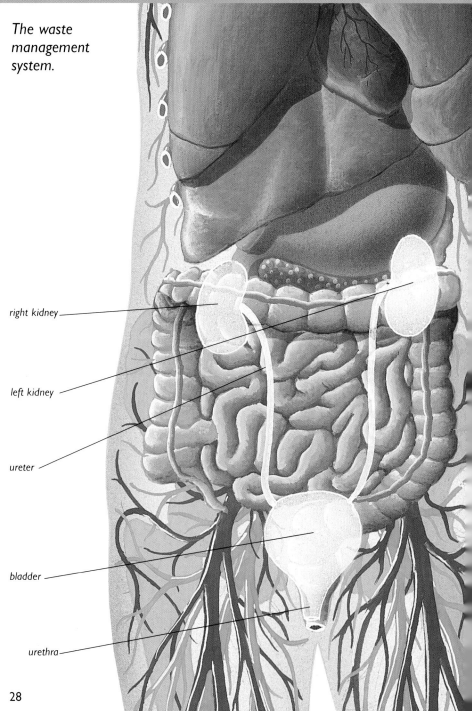

right kidney

left kidney

ureter

bladder

urethra

Our body produces many different types of waste products, including urine and feces.

Bladder

The two kidneys help to extract any excess water and waste from the blood, which is then passed into the bladder via the ureters.

When we want to empty our bladder, it contracts, sending the urine out of our body through the urethra. On average, people empty their bladders six to eight times a day, excreting four to eight cups of fluid.

The adult bladder is able to hold about two cups of urine.

Rectum

Feces are stored in the rectum, and contain undigested food such as fiber, dead cells, bacteria, bile, and water.

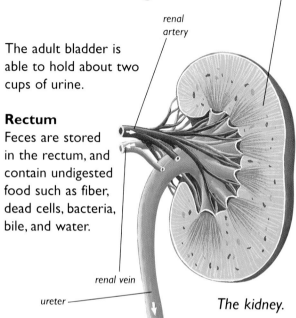

renal capsule

renal artery

renal vein

ureter

The kidney.

Body Facts

• It takes up to 30 hours for food we have eaten to pass out of our body.

• You need to drink 50-70 ounces of fluid every day for your kidneys, bladder, and bowel to work efficiently.

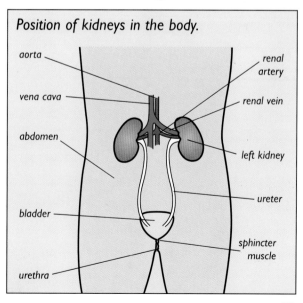

Position of kidneys in the body.

aorta

vena cava

abdomen

bladder

urethra

renal artery

renal vein

left kidney

ureter

sphincter muscle

29

Hands

The hand allows us to grip, touch, and convey words, such as "hello," by waving.

Bones

The hand is made up of 27 bones. Eight bones (the carpus) form the wrist joint, and two bones (the radius and ulna) make up the forearm.

Five bones (the metacarpals) form the palm of the hand. The remaining fourteen bones (the phalanges) form the fingers and thumb.

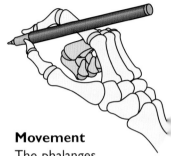

Movement

The phalanges allow four movements to occur in the fingers—flexion (bending), extension (straightening), abduction (movement outwards), and adduction (movement inwards).

Thumbs

Due to the position and shape of the joint, the thumb is able to move in a circular pattern. This allows us to perform complex activities, such as gripping a pen, or opening a jar. Try writing without using your thumb—you'll soon see how useful it is!

Bones in the hand.

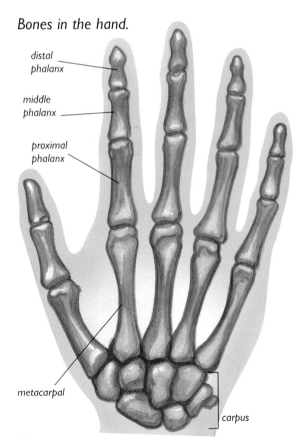

distal phalanx

middle phalanx

proximal phalanx

metacarpal

carpus

Feet allow us to stand, walk, and run on flat or uneven surfaces. The foot and ankle are supported by several strong ligaments, which provide stability and prevent the foot from rolling over.

Bones in the feet.

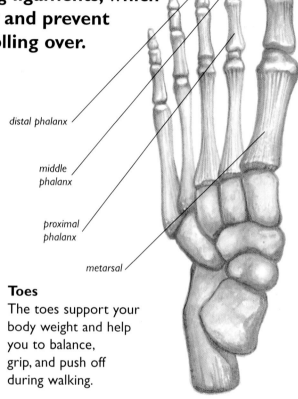

distal phalanx

middle phalanx

proximal phalanx

metarsal

Bones
The foot is made up of 26 bones. Seven bones form the ankle joint (the tarsus) along with two bones in the lower leg, called the tibia and fibula. Like the hand, there are five metatarsal bones and fourteen phalanges that form the toes. The heel and ankle support most of your body's weight.

Arches
There are three arches in the foot. They provide shock absorption and assist in balance. Without these arches, walking would be very uncomfortable, and running would be impossible.

Toes
The toes support your body weight and help you to balance, grip, and push off during walking.

Body Facts
- Each foot is able to take up to five times your body weight when running.
- The Achilles tendon, which runs over the ankle and foot, is the strongest tendon in the human body.
- The sole of the foot contains the thickest skin in the body.

We all began life as a single fertilized cell, called an embryo.

Weeks 0-13 (First Trimester)

Fertilization occurs when the male sperm penetrates the female ovum (egg). The embryo divides every few hours until there are thousands of cells, which start to form complex body structures, such as the heart and blood system.

Seven days after fertilization, the embryo implants onto the wall of the uterus (womb), which will be its home for the next nine months. By six weeks, the heart and circulatory system have formed. By eight weeks, the embryo is known as a fetus, and is beginning to look like a human. By twelve weeks, all the major body organs have formed.

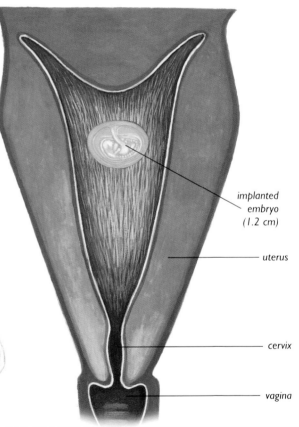

implanted embryo (1.2 cm)

uterus

cervix

vagina

The fetus at eight weeks.

Body Facts
- By three months, the fetus will already have its own individual fingerprints.

Weeks 14-26 (Second Trimester)

The placenta, (the baby's life support system), allows nutrients from the mother's blood to be passed to the baby. The baby's bones are forming rapidly, as are the nails.

At fifteen weeks, a fetus can move its head and limbs, grip, and even suck its thumb.

At twenty weeks, the eyes remain closed, although the eyelids and eyebrows have formed.

At 24 weeks, the baby weighs less than 3 pounds, and measures around 10 inches long.

By the end of this trimester, the baby is able to hear, and will respond to voices and sounds from outside the womb.

The fetus at 24 weeks.

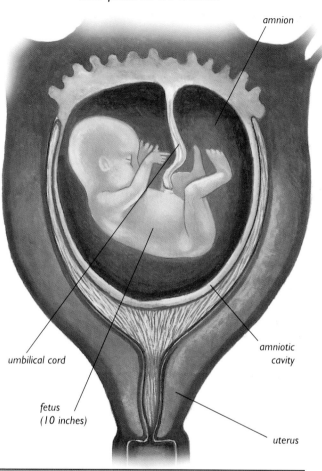

amnion

umbilical cord

amniotic cavity

fetus (10 inches)

uterus

Body Facts

- Most mothers will first be able to feel their babies move between 16 and 21 weeks.

- A baby's heart beats approximately 140 times per minute, twice the speed of an adult.

- Research has shown that babies who are played a specific type of music while inside the womb will recognize it once born.

The fetus at 38 weeks.

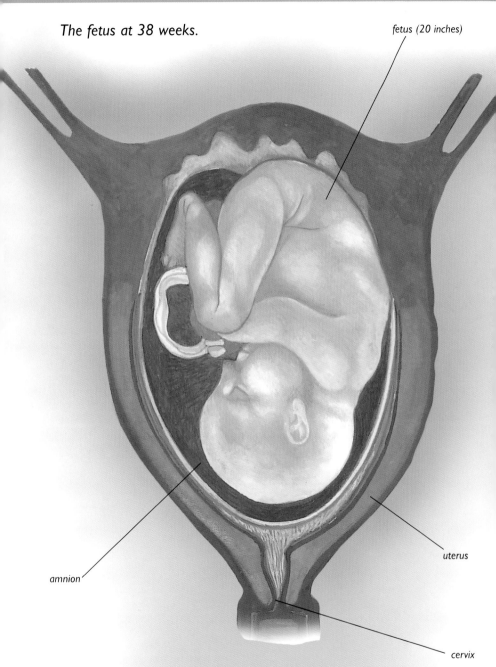

fetus (20 inches)

uterus

amnion

cervix

An ultrasound photograph of a fetus.

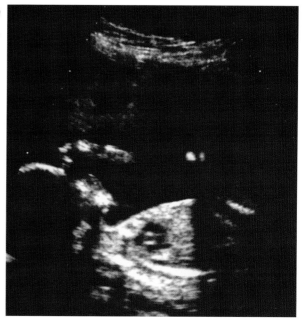

Weeks 27-40 (Third Trimester)

During the last trimester, the baby will grow up to half an inch a week, and will be storing fat under the skin in preparation for birth.

At 29 weeks, a baby can open and close its eyes, and will often have a head of hair.

At 36 weeks, a baby's lungs are ready to take their first breath.

Most babies are born between 38 and 42 weeks, weighing 6-8 pounds and measuring 20 inches long.

During the third trimester, the body will have grown to a stage where it is in proportion with the head, which grows much more quickly.

The Mother

To make sure the baby is born healthy, it is very important that the mother looks after her own health carefully. She shouldn't smoke, drink, or take drugs.

Birth

When the baby is ready to be born, the mother will have contractions, and the cervix will dilate in preparation for the birth.

Body Facts

- Most babies are born with blue eyes, but the color can change after six weeks.

- Did you know that a Caesarean section is said to be named after the Roman emperor Julius Caesar, (100-44 BC) who was born in this way?

⬤ **Birth to Toddler**

The first few years of life involve the development and acquisition of many new skills. During this phase, the human body will undergo many changes.

0-3 Months

On average, babies gain 7 ounces a week for the first three months of life, and by the end of this stage will have doubled their birth weight. Babies are born with a sucking reflex, enabling them to drink milk from the mother's breast or a bottle.

Milk is the only source of food for the first few months of a baby's life.

Just born.

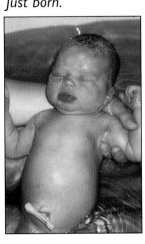

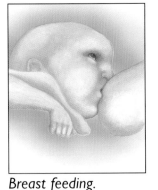

Breast feeding.

At three to four weeks old, a baby will be able to support its head for a few seconds. It is not until it is three months old that it will be able to fully support its head.

A baby supporting its own head.

At around six weeks old, a baby will be able to smile. All babies appear to smile before this stage, but it is as a result of gas, or filling their diaper!

First smile.

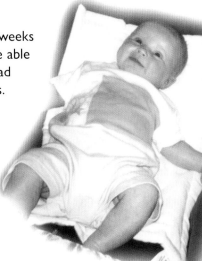

3-6 Months

From three to six months old babies gain, on average, six ounces a week, and will grow by an inch a month. A baby should be given solid food, such as baby rice and fruit, at around sixteen weeks. Milk still forms an essential part of the diet.

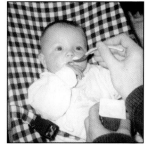

Taking solid food.

Teething can begin at around four months, and the first baby tooth may be visible by five to six months.

6-12 Months

From six to twelve months, growth slows down to two ounces a week and around half an inch in length a month. The diet is now becoming more varied in texture and taste, and babies can self-feed. Milk still forms a large part of the diet.

From six to eight months old, a baby will be able to sit independently. By nine months, it may begin crawling, and can stand from ten months. Some babies start to walk before their first birthday.

First words can be spoken from ten months old.

1-3 Years

The brain is still developing rapidly, allowing for complex tasks such as counting, reading, and writing. Most children will be walking by eighteen months old. At two years old, a child is very stable on its feet and can even run. Speech will improve daily.

Body Facts

• Newborn babies are only able to focus on objects up to twelve inches away. They will respond to changes in brightness and movement, but are unable to focus.

• Babies are attracted to black and white patterns, as opposed to colored ones.

Puberty is the time when our reproductive organs are stimulated to develop by hormones and reach maturity.

Average Age
The average age for puberty to begin is ten to twelve years for girls, and twelve to fourteen years for boys.

Female Changes
Changes in the female body include the release of eggs from the ovaries, and the beginning of the menstrual cycle (the period). The breasts develop, pubic and underarm hair grow, the pelvis widens, and the skin becomes oily and prone to spots. The body will also store fat, particularly on the breasts, hips, and thighs. The body is now ready to carry a baby.

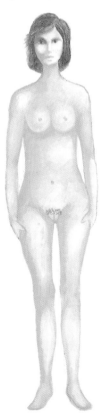
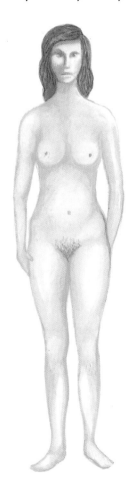

Male Changes

Changes in the male body include a rapid growth spurt, the voice box getting larger, and the voice deepening. There will be hair growth on the face, chest, tummy, underarms, and pubis, sperm will be produced, and the skin will thicken and become oily.

Body Facts

- The testes produce around 500 million sperm every day!

- A sperm measures 0.002 in. long, so is undetectable by the human eye.

- At birth, a female has 600,000 immature eggs present in the ovaries.

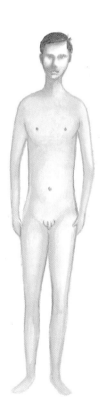

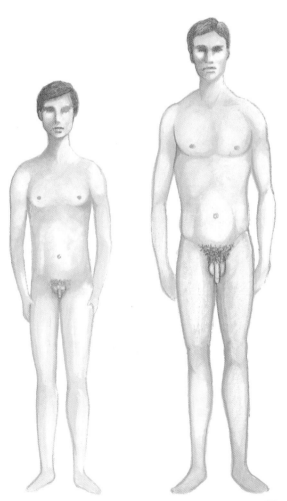

We need lots of different types of food for our bodies to work. There are five main groups of foods that our diet should contain: proteins, carbohydrates, fats, vitamins, and minerals.

Carbohydrates

Carbohydrates provide us with most of our energy in the form of sugars and starch. Eaten in excess, carbohydrates are turned into fat and stored. Carbohydrates include bread, sugar, breakfast cereals, rice, and potatoes.

Minerals

We need small amounts of minerals for all our body organs to work efficiently. A mixed diet should contain all of the minerals we need. Examples of minerals are calcium, iron, phosphorus, iodine, sodium, and potassium.

Vitamins

Vitamins are needed for our bodies to be healthy. We can not make our own vitamins, therefore they must come from what we eat. A mixed diet should contain all of the vitamins we need. Vitamins are either water- or fat-soluble. Water-soluble vitamins (B and C) are damaged by heat, and foods containing these, such as vegetables, should not be overcooked. Fat-soluble vitamins (A, D, E and K) can be stored in the body and should not be eaten to excess.

You Are What You Eat

Proteins

Proteins help us to make new cells and tissues. They are made up of amino acids, some of which our bodies can produce, and others that we can only get from certain foods. We are unable to store proteins in our bodies, so we should try to eat some every day. Proteins include meat, fish, eggs, milk, cheese, pulses, and rice.

Fats

Fats give us energy and can be stored within the body. There are two types of fats—saturated and unsaturated. Saturated fat is derived from animal products and processed foods such as meat, butter, milk, eggs, cheese, chocolate, and crisps. Unsaturated fat is derived from vegetable and fish sources such as olive oil, sunflower oil, and mackerel. We should try to eat more unsaturated fat than saturated fat, as too much saturated fat has been linked with heart disease.

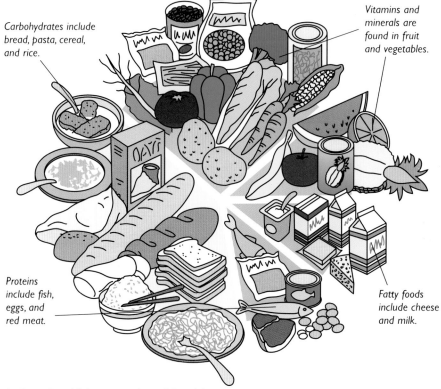

Carbohydrates include bread, pasta, cereal, and rice.

Vitamins and minerals are found in fruit and vegetables.

Proteins include fish, eggs, and red meat.

Fatty foods include cheese and milk.

A diet should be varied and healthy.

Vitamins

Vitamin	Source	Function
A	Dairy produce, liver, fish oil, carrots, red peppers, tomatoes.	Aids growth of tissue and body structures, helps fight infection, aids vision, healthy skin, and strong bones.
B	Meat, dairy produce, whole grain cereals, yeast extract, bananas.	Promotes growth, healthy nervous system, aids digestion, blood cell formation, helps fight infection.
C	Citrus fruits, green vegetables, potatoes.	Promotes growth, tissue repair, healthy skin, aids iron absorption.
D	Dairy produce, fish.	Helps absorb calcium and phosphorus, needed for strong bones.
E	Vegetables, oils, nuts, whole grain cereals, dairy produce.	Red blood cell formation and maintenance of blood cells.
K	Fish, green vegetables, fruit.	Blood clotting.

Minerals

Mineral	Source	Function
Calcium	Dairy produce, green vegetables, nuts, fish.	Health and formation of bones and teeth, blood clotting, muscle contraction.
Phosphorus	Cheese, oatmeal, liver, kidney.	Hardening of bones and teeth.
Sodium	Fish, meat, dairy produce, bread, salt.	Supports the body's immune system.
Potassium	Fruit, vegetables.	Muscle contraction, transmission of nerve impulses.
Iodine	Liver, kidneys, eggs, green vegetables.	Essential in production of thyroxin.
Zinc	Shellfish, meat, nuts, cereals.	Supports the body's immune system.
Iron	Red meat, pulses, cereals, green vegetables, dried fruit.	Hemoglobin formation in red blood cells.

We are all aware of the importance of eating the correct diet and exercising to ensure we maintain a healthy body. Here are a few tips to help you on your way.

Exercise

Aim to exercise for thirty minutes, two to three times a week. Ideally, choose a form of cardiovascular exercise that will increase your heart rate. This includes walking (quickly), running, playing football, swimming, and cycling. Why not try walking to school as part of your exercise regime?

Fruit and Veg

Try to eat five portions of fruit and vegetables every day, including fruit juice and dried fruit.

Fruit and vegetables are high in vitamins and minerals that are essential for a healthy body and immune system.

Junk Food

Cut down on the amount of junk foods you are eating, as it is high in saturated fats. This type of fat is bad for the heart if eaten in large quantities. Avoid eating or drinking foods that

contain a lot of sugar. Excess sugar will be stored by the body as fat. It is also bad for your teeth.

Save fast food, chocolate, and chips for occasional treats.

Brush Your Teeth

Brush your teeth twice a day to remove food, acids, and plaque.

Visit your dentist for regular checkups, so that any possible problems can be sorted out early. You only get one set of adult teeth, so look after them!

Sleep

Aim to get at least eight hours of sleep a night. We need to sleep to allow our body to recover from the day's activities, and in order to produce new cells and aid healing.

A lack of sleep will make you irritable, prone to illness, forgetful, and tired!

Glossary

Adam's apple the adult male voice box.

Alveolus a tiny air sac in the lungs where oxygen and carbon dioxide are exchanged.

Artery a large blood vessel that carries oxygenated blood from the heart to the tissues.

Atrium the atrium is comprised of the two chambers of the heart: the left and right atria.

Auditory nerve carries sound waves from the cochlea (inner ear) to the brain.

Automatic nervous system part of the nervous system that controls automatic functions such as coughing, breathing, and heartbeat.

Axon a long fiber that runs from the nerve cell body.

Bacteria single-celled organism, that can be good or bad.

Bile waste product produced by the liver.

Blood clotting when parts of the blood stick together and turn solid to stop bleeding.

Bone marrow the center of the bone. Produces red blood cells, some white cells, and platelets.

Brain stem the bottom of the brain connecting the brain and spinal cord, where automatic functions are controlled.

Bronchiole the bronchi divide into a network of tiny tubes, called bronchioles, that allow air to move in and out of the lungs.

Bronchus the right and left bronchi branch out from the trachea, allowing air to pass into the lungs.

Caesarean section an operation performed to remove a baby directly from the womb, rather than through the vagina.

Capillaries tiny blood vessels that run between arteries and veins.

Carbohydrate one of the main food groups, made up of sugars and starch.

Carbon dioxide the waste product of respiration, which is breathed out.

Cardiac muscle special muscle that forms the heart.

Carpals eight bones that help form the wrist.

Cartilage a gristly, tough material that lines the ends of bones where they form joints, preventing wear and tear to the bone.

Central nervous system consists of the brain and spinal cord.

Cerebellum found beneath the cerebrum. It helps control balance and co-ordinate movement.

Cerebrum the largest part of the brain, it is divided into two halves, called the cerebral hemispheres.

Circulation blood flow around the body.

Cochlea part of the inner ear, where sound waves are transmitted via the cochlear nerve to the brain.

Collagen a protein found in connective tissue, such as tendons and skin.

Cornea the clear front of the eye that helps focus light.

Dermis inner layer of skin, beneath the epidermis.

Dendrites finger-like extensions at the end of a neuron, which receive signals and carry the message to the nerve cell body.

Digestion the process of converting food into energy.

Embryo a fertilized egg is called an embryo, which develops over three months to form a fetus.

Epidermis the outer layer of skin.

Enzyme a chemical that speeds up the breakdown of food in digestion.

Esophagus a tube (the gullet) running from the mouth to the stomach.

Expiration breathing out.

Excretion discharging waste products from the body.

Fat one of our main food groups that are high in energy and can be stored in the body.

Feces waste product of digestion that we excrete.

Fertilization when a sperm enters an egg, forming an embryo.

Fetus develops from an embryo after three months of pregnancy.

Fibula a thin, long bone on the outer edge of the lower leg.

Gall bladder an organ attached to the liver, where bile is collected and enzymes are released.

Gluteus maximus a large powerful muscle that helps to form the buttocks.

Hemoglobin oxygen-carrying pigment found in red blood cells.

Hormone a chemical controlling many body processes and affecting specific body structures.

Inorganic a substance that is not made up of living components.

Inspiration breathing in.

Intestine comprises of the small and large intestine. Absorbs nutrients and forms waste for excretion.

Keratin a protein found in hair and nails.

Larynx voice box.

Lens helps focus light onto the retina. Forms the colored part of the eye.

Ligament a strong fibrous tissue that connects bone to bone.

Liver a large organ in the body that helps break down food.

Mandible the jaw bone.

Menstrual cycle the uterus lining breaks down and bleeding occurs for a few days every month— the period.

Metacarpals bones that form the palm of the hand.

Metatarsals bones that form part of the foot.

Mineral one of the main food groups needed for your body to function efficiently.

Nerves bundles of fibers that carry messages to and from the brain.

Neuron a single nerve cell that sends and receives messages from the brain.

Olfactory nerve carries smell signals to the brain.

Ossicles small bones in the middle ear that help carry sound waves to the eardrum.

Ovaries the female reproductive organs, which release and store eggs.

Ovum a single egg.

Oxygen a gas found in air that is essential for breathing and human life.

Palate the roof of the mouth.

Pancreas an organ that secretes enzymes to aid in digestion and produces the hormone insulin.

Peripheral nervous system made up of 31 pairs of spinal nerves extending from the spinal cord to all areas of the human body.

Phalanges bones that form the fingers and toes.

Pharynx the throat.

Pinna the external ear.

Placenta allows nutrients to be passed from the mother to the baby during pregnancy.

Plasma the liquid in which blood cells and platelets float.

Platelets cells which help blood to clot.

Protein essential food group. Proteins help build body structures.

Puberty stage where male and female reproductive organs mature.

Pupil the center of the eye, used to channel light.

Radius a long bone in the forearm connecting the elbow and wrist.

Rectum where waste products from the intestine are excreted.

Reflex an automatic response by the body to a stimulus.

Receptor cell type of cell that receives messages concerning body sensations such as smell.

Glossary

Red blood cell carries oxygen around the body.

Retina the back of the eye where images are interpreted.

Respiration process of gas exchange allowing humans to live. Oxygen is absorbed on inspiration by the body tissues, and carbon dioxide is removed on expiration.

Ribcage bones in the chest that help protect vital organs, such as the heart and lungs.

Saliva fluid present in the mouth that helps break down and soften food.

Scalp the skin covering the top of the head, where hair grows.

Skeletal muscle allows joints to move.

Skull bones that form the head.

Smooth muscle lines most of the internal organs.

Sperm produced by the male in the testes. Combines with the female egg to produce an embryo.

Spine forms the backbone, supporting the head, and protecting the spinal cord.

Spinal cord links the brain to the rest of the body via spinal nerves that branch from the spinal cord.

Stapes the stirrup bone, which forms part of the middle ear.

Sternum forms the breastbone and provides attachment for the ribs.

Sweat glands lie at the surface of the skin and help to cool down the body.

Tarsus bones that help form the ankle joint.

Taste bud a type of receptor cell, found in the mouth, which detects taste.

Tendon a strong, fibrous tissue that connects muscle to bone.

Testes the male reproductive organs that produce sperm.

Thyroxin a hormone produced by the thyroid gland that affects most of your body's organs and tissues.

Tibia a long bone in the lower leg connecting the knee to the foot, and helping to form the ankle joints.

Tissue a large number of cells with a similar structure and function.

Trachea the windpipe. Connects the throat to the lungs.

Trimester a stage of pregnancy lasting approximately three months. There are three trimesters in pregnancy.

Ulna a bone in the forearm that connects the elbow to the wrist.

Ureters tubes that carry urine from the kidneys to the bladder.

Urethra a tube that carries urine from the bladder out of the body.

Urine a waste product produced in the kidneys.

Vasoconstriction narrowing of the blood vessels.

Vasodilation widening of the blood vessels.

Vein a blood vessel that carries deoxygenated blood from the tissues back to the heart.

Ventricles chambers in the heart that pump blood to the lungs and around the body.

Vitamins essential part of any diet, ensuring the human body runs efficiently and stays healthy.

White blood cell helps to fight infection.